EARLY GEOLOGIC TRAINING OF THE APOLLO ASTRONAUTS

A personal recollection

by

Bob Regan

APOLOGIA

This text is quite different from the other books I have authored as it is mostly a personal recollection. This form is necessitated by the fact that during the time of these events I kept no notes. Although there are scores of material, including photos and movies, in the USGS and NASA archives I did not access any but relied on what was readily publicly available and what personal photos and memories survived over the years.

Please excuse the many instances of the use, throughout the text, of the first person singular.

PREFACE

The U.S. Race to the Moon was not just to land a man on the Moon and successfully return him to Earth, but also to explore and examine the Moon to determine its composition, structure and ultimately, origin. Consequently, the lunar astronauts were trained not only on their spacecraft and lunar flights but also on the earth sciences of geology and geophysics.

The geologic and geophysical training of the lunar astronauts was a project of the U.S. Geological Survey (USGS) and for four years I was a member of that project. There has never been a well-documented, official report or manuscript on the geologic training program. Years ago one was being prepared but abandoned after the death of the principal author, Dale Jackson. However, the excellent book by Don Wilhelms, *To a Rocky Moon*, affords a geologist's view of the entire lunar program including salient aspects of the geology training program.

This text offers my recollections of this truly amazing program and hopefully also offers some additional perspective on this aspect of the Country's lunar program. It does not offer a detailed geological viewpoint but hopefully enough of one so that what was done and why can be fully appreciated.

INTRODUCTION

The geologic training of the lunar astronauts began as a USGS project at Ellington Air Force Base, the forerunner of the Johnson Manned Spacecraft Center, in Houston, Texas. Soon after the inception of the project NASA hired several geologists to complement the work of the USGS personnel. This led to some disagreements in the new project especially as the USGS project leader, Dale Jackson, was quite contentious.

In order to ease tensions between the two agencies Dale was replaced by Al Chidester and eventually the project relocated to the USGS Branch of Astrogeology in Flagstaff, Arizona. The geologic setting of Flagstaff was the main reason that the Branch of Astrogeology was located there. The City is surrounded by various volcanic features including cinder cones and lava fields and Meteor Crater was only thirty miles away. Such features were significant at a time when a great debate existed over the volcanic or meteoritic origin of the Moon's craters.

Another reason for the Branch's location in Flagstaff was the excellent atmospheric conditions which combined with little ambient light afforded ideal conditions for telescopic observations of the Moon. The geologic mapping of the Moon, via telescope, was one of the prime functions of the Branch of Astrogeology.

I was finishing my M.S. graduate studies in geophysics at Boston College's Weston Observatory when a fellow graduate student mentioned that Marty Kane, a geophysicist with the USGS, had contacted Father Skeehan, head of the Geology Department. Marty was looking for a geophysicist to come to work at the USGS in Flagstaff, Arizona. He was leaving for several years to attend graduate school at St. Louis University to obtain his Ph.D. under the government sponsored 'In Training' program.

My fellow graduate student, a gregarious bachelor, did not relish leaving the urban setting of Boston for a small city in Arizona. I, however, was getting married and thought this would be a good entry into the field of geophysics and an exciting opportunity. I applied for the position and was ultimately accepted. Several years later I was to follow in Marty's footsteps as I was afforded the opportunity to obtain my Ph.D. under the same 'In Training' program.

After an interesting drive on Route 66 my wife and I finally arrived in Flagstaff in late December, 1964. It was a welcome relief as only thirty miles away from Flagstaff it was cloudy and we could not see the mountains we knew to be there from pictures that had been sent to us. The only scenery at that point was miles of high desert flat land. But the mountains were there and the small City of Flagstaff was indeed quite picturesque. At an elevation of 7,200 feet it is quite different from the standard images of the more populous locales of southern Arizona. Although it was a city it seemed like a small town especially when compared to Boston.

The 7,200 foot altitude required some acclimation, especially for my wife who was several months pregnant at the time. We later learned that while my body had to adjust for only itself, she had to adjust to the altitude for two beings.

The Branch of Astrogeology was relatively new having been started by Gene Shoemaker only a year or two before my arrival. Gene was a remarkable man and scientist and it was a delight to have been involved with him. A complete history of the Branch can be found in the USGS Open-File Report by Shaber and the book *To a Rocky Moon* by Wilhelms.

The Branch was divided into two components entitled Manned and Unmanned Lunar Exploration. As the title suggested the group in Unmanned Lunar Exploration focused on telescopic geologic mapping of the Moon as well as mission planning for the Ranger and Surveyor spacecraft and analyses of their data. The Manned Lunar Exploration division, where the Astronaut Training Project was located, worked on designing and planning methods and techniques for manned geologic exploration of the Moon.

The personnel in both groups, as well as those in the various support groups, were extraordinary. All were young, talented and committed to the goal of unraveling the mysteries of our nearest astronomical neighbor. It was a little intimidating for me, a recent graduate student, to be cast among this pack of scientists. One example of the young scientists there was my office mate when I first started work; a Ph.D. from Harvard named Jack Schmidt who would years later walk on the surface of the Moon during the Apollo 17 mission.

Naturally when I arrived, my boss, the head of the Astronaut Training project, Al Chidester, was out of town. However, he eventually returned and I began my efforts under his direction. Simply put, Al was the ideal person for the position of head of the Astronaut Training project. As I witnessed over the years he blended well with the NASA geologists, got along famously with the astronauts, and handled all aspects of the project's activities with diligence. I doubt the project, which involved so much interaction with NASA, the astronauts, various earth scientists, and dignitaries would have succeeded if not for Al's personality, competence and demeanor.

Meteor Crater: Gene Shoemaker and seated with glasses, Al Chidester

ASTRONAUT TRAINING PROJECT

From late 1964 to 1967, I worked as a geophysicist in the Astronaut Training Project. Our job was the early, general geological/geophysical training of the Apollo astronauts. This training was carried out through means of classroom lectures and field exercises. Later training, which followed, was mission specific training.

The mainstay of the project was the geologic field trips. Again, at that time, there was a great debate over the origin of the craters on the lunar surface, i.e., whether they were of volcanic or meteoritic origin. Thus, one goal of the project was to expose the astronauts to geologic examples of various craters around the world, i.e., lunar analogues. Of course such trips were often backstopped by excursions to the various volcanic features in the vicinity of Flagstaff and nearby Meteor Crater.

Al Chidester set up all the field trips. He certainly knew the geology of each area but he also knew experts in the geology of the areas and invited them to lead the trips. These experts were not always from the USGS and this method of conducting the field trips was instrumental in diffusing any rivalry with the NASA geologists and helped ensure the success of the program.

The field trips were far from clandestine as every field trip involved not only the astronauts, but also geologists from NASA and the Branch of Astrogeology, NASA Training and Public Relations personnel, and USGS and NASA still and motion picture photographers plus several people associated with the locale. If there were five astronauts then the entourage totaled about thirty people. In addition the fact that astronauts were coming to a locale always leaked out and we had to deal with local news reporters.

Initially the field trips involved geological lectures and guided field mapping of the areas. Later trips involved the astronauts using transmitting microphones, and recording equipment, so they could practice verbal geologic observations and mapping as they would be doing on the lunar surface.

In addition to support activities on all field trips I also provided field geophysical training at several sites, i.e., Nevada Test Site, Meteor Crater, and Zuni Salt Lake. Again, the goal was to demonstrate the differences between geophysical signatures of volcanic and meteoritic craters. The astronauts particularly liked the geophysical training as the equipment involved switches and dials with which they could easily relate.

Classroom Lectures

Prior to the start of this phase of the training the astronauts had received several lectures by USGS and NASA geologists on the basics of terrestrial and lunar geology and also basic mineralogy and petrology. For my part, during our part of the training, I can recall only two classroom lectures that I was involved in.

The first was a lecture on basic exploration geophysics by Marty Kane and me. This, like all the lectures, was held at Ellington Air Force Base in Houston. I only recall that the lecture went well and was well received. I also recall thinking about the composition of the class. I doubted that I would ever again face such a formidable group of students.

The second lecture was to be a discussion (debate?) between Gene Shoemaker and someone else, whose name I cannot recall, on meteoritic versus volcanic origin of the Lunar craters. Gene had to cancel and Hal Mazursky was drafted to replace him. Hal worked in the Menlo Park, California offices of the Branch of Astrogeology and I had never met him.

I arrived in Houston to find Hal, my roommate, already in the hotel room. He struck me as a long haired, intense, but delightful character. His *ex temporaneous* presentation the next day was brilliant. He effectively argued for the meteoritic origin of lunar craters citing ejecta blankets, overlapping craters, etc. Later Hal relocated to Flagstaff and it was delightful, not to mention entertaining, to have such a talented individual in the same office.

Did I mention that Hal was intense and dedicated? One story can illustrate this. I was at a meeting with Hal in the San Francisco area. Around 9:30 am, Hal excused himself and didn't reappear back at the meeting until 3:30 pm. On the trip back to Flagstaff I learned from him that his absence from the meeting was due to a trip to Houston where he argued for a window to be placed in the Skylab spacecraft.

Geologic Field Trips

The main reason for the geologic field trips was to study different types of craters, lunar analogues. In addition to trips around Flagstaff, the field trips included visits to: Zuni Salt Lake, New Mexico; Grand Canyon (a basic geology trip); Katmai, Alaska; Hawaii; Iceland; Pinacate Volcanic Field, Mexico; Medicine Lake, California; Cimarron, New Mexico; Nevada Test Site, Nevada; Valles Caldera, New Mexico; Bend, Oregon; and Marathon Basin, Texas.

The geologic training of the astronauts was only one aspect of the intensive, diverse, wide-ranging continual training that they were involved in. Consequently, scheduling of any aspect of the geologic training was sometimes problematic and it was rare to have all the astronauts available for training at the same time. This was especially true for the geologic field trips as each included several days of travel and training. Thus, many sites were visited several times with different groups of astronauts.

Nevada Test Site

The first field trip I was involved in, actually a series of three within a few months, was to the Department of Energy's Nevada Test Site, now known as The Nevada National Security Site. The astronauts participating in these trips were: Buzz Aldrin, Neil Armstrong, Charles Bassett, Allan Bean, Eugene Cernan, Michael Collins, Walt Cunningham, Donald Eisele, Dick Gordon, Rusty Schweikart, Dave Scott, Roger Chaffee, and Elliott See.

The Nevada Test Site, located about 65 miles Northwest of Las Vegas was an ideal place to study impact like craters as well as an area to demonstrate various geophysical surveying techniques such as seismic exploration. Located in the Basin and Range province, it has deep sedimentary filled valleys and also surrounding volcanic mountains. The field trip leaders for these trips were Will Carr and Bob Christiansen, both of the USGS

The craters studied were produced by underground nuclear explosions during the Operation Plowshare Program which was designed to show peaceful uses for nuclear explosions such as excavations for harbors, canals, and open pit mining. Of particular interest was the Sudan Crater. A 104 kiloton nuclear explosion on July 6, 1962 created the 1,280 foot wide, 330 foot deep, crater. The crater which is a man-made object that can be seen from space has now been placed on the National Register of Historic Places. Sudan offered an opportunity to examine an impact like crater and ejecta that was formed in unconsolidated sediments.

My particular involvement in this trip was to demonstrate the seismic exploration technique, wherein small subsurface explosives are detonated and the shock wave recorded on devices called geophones in well surveyed locations. Interpretation of the recorded results can provide information on subsurface structure.

To prepare for the seismic demonstration, we arrived well ahead of the astronauts to perform surveys in different areas to find an optimal one for demonstration. Initially, we were in an area that had remnants of buildings that had been destroyed by above ground nuclear explosions. After working several days in this area we were told that NASA considered the area too hot (radiologically) for the astronauts. That made us all feel a little like guinea pigs.

At any event I set off cross-country to find another locale. After travelling overland for a while I came to a cluster of house trailers and vehicles and was met by an armed guard. He asked what I was doing there and I explained. He then told me that I had stumbled into the advance area for an underground nuclear explosion and since it was soon scheduled I would not be allowed to leave. He then pointed out a tower a short distance away and said that was where the shot was to take place. He further explained that the tower was designed not to be affected by the blast so that after the shot they could re-drill into the blast site.

A short while later I noticed everyone, including the guard, donning radiological protection suits. That is everyone but me. I turned the jeep around so that it was headed back the way I had come and left the motor running. The guard nodded O.K.

There was a countdown and a large 'thump'. We were in unconsolidated sediments and I witnessed ground waves about 3-4 feet in amplitude coming toward us. At one point I was looking at the tower over a house trailer and then my view of the tower was underneath the trailer. The tower fell. I looked to the guard. He nodded and I took off.

There is one more salient recollection from this trip. I still recall one evening walking in the desert with Roger Chaffee. After a while I asked him if he was ever afraid during space flight. He said No and that the only place he worried was sitting on top of the rocket at the cape as he had no control. How prophetic as this wonderful person would perish on top of a rocket at the cape. He was a victim, along with Ed White and Gus Grissom, of the Apollo 1 fire on January 27, 1967.

Meteor Crater, Arizona

In April of that year (1965) our next trip was to Meteor Crater. The astronauts were: William Anders, Charles Bassett, Eugene Cernan, Roger Chaffee, Michael Collins, Walt Cunningham, Donald Eisele, Dick Gordon, Rusty Schweikart, Dave Scott, and C.C. Williams.

Meteor Crater, located about 30 miles east of Flagstaff, is 3,900 feet in diameter and 570 feet deep. It is surrounded by a rim that rises 149 feet above the surrounding plains. Because of the high rim, from a distance it look like a mesa or butte and was once known as Coon Butte.
The seminal work by Gene Shoemaker in the early 60s established that the crater was the result of meteorite impact and this work is considered the first definitive proof of meteorite impact on Earth. Gene was fond of saying that it was a relatively small, slow moving meteorite traveling at a low angle that caused the crater. As he often said it was more the Earth hitting the meteorite than the other way around. The crater occurs in a region of well layered sedimentary rock and the rock layer sequence is inverted in the rim. It is like the rocks having been folded back upon themselves.

Prior to and after the training trip the Branch of Astrogeology conducted a full geophysical investigation of the crater to document the subsurface structure. This work has since been published in the Journal of Geophysical Research.

This trip was the astronaut's first experience with a meteorite crater. Gene Shoemaker provided a half day tour of the crater and the rim and then the astronauts set out to study and map the various features. At the bottom of the crater I held another demonstration of the seismic survey technique which showed the structure of the crater bottom.

I had been working in the crater bottom for several weeks and felt in good shape having to climb in and out of the crater each day. So at the end of one day I challenged the astronauts to a race out of the crater. What an upcoming as to no one's surprise I finished last.

Meteor Crater: Far left with back to camera, Mike Collins, Ray Zedekar (NASA), RogerChaffe, Dave Scott, Jack Schmidt, Bob Regan , Leaning on water can, Charlie Basset, Uel Clanton (NASA):

Meteor Crater: Geologic lectures on the crater rim

Meteor Crater: Dick Gordon, Bob Regan, C.C. Williams

Meteor Crater: Bob Regan lecturing on results of geophysical survey.

<u>Katmai, Alaska</u>

In late June, 1965 we had a training trip in the Katmai National Park and Preserve in Alaska. The astronauts were; Buzz Aldrin, William Anders, Charles Bassett, Allan Bean, Eugene Cernan, Roger Chaffee, Walt Cunningham, Rusty Schweikart, Dave Scott, and C.C. Williams. The leader of the trip was Bob Smith (USGS)

The June, 1912 eruption of Novarupta Volcano deposited silica-rich pumice and ash over a large area in the Valley of Ten Thousand Smokes. At the end, 40 square miles of land was covered with as much as 700 feet of volcanic deposits. Thousands of small holes and cracks developed in the ash deposits allowing gas and steam from the heated ground water to escape; thus, the name of The Valley of 10,000 Smokes. Following the eruption the summit of the volcano collapsed by 1500 feet. The volcanic features in this area were almost pristine so it afforded a good training area to study volcanics.

I was delayed going to the field trip by the birth of my first daughter. One of the USGS photographers, Jim McCord, was also delayed and we traveled together to Anchorage. I don't recall how we got from Anchorage to a rustic guest house where we stayed for an evening. The remote place contained a true sauna and Jim and I sauned complete with a short dash to a cool lake. In the morning a helicopter from the military's Air-Sea rescue group landed to carry us to Katmai. What a trip. We flew close to the ground with Jim and I secured in the open door spotting the wildlife and terrain below. We were delivered to the Brooks Lake Lodge in the Park where we joined the rest of the group. We were also joined by several other helicopters that would offer us transportation to The Valley of 10,000 Smokes throughout the field trip.

The Brooks Lake Lodge is a high-end facility and we fully occupied it. It was rustic yet elegant and undoubtedly the best field trip quarters we were to ever enjoy. Each day the helicopters would transfer us to the Valley. The first day or so Bob Smith provided an excellent geologic view of the various volcanic features. Then, for the first time, we employed transmitting microphones, receivers and recorders. The astronauts were paired up, equipped with transmitting microphones, and each pair worked on various volcanic features providing a verbal mapping of what they were encountering. This was the way they would make any geological observations on the moon and this was their first experience in doing so. On the receiving end for each pair was a geologist who made notes and during debriefing offered a critique of their observations.

I followed Walt Cunningham and Alan Bean and it was an enjoyable experience. They did a competent job of describing the geology and their geologic descriptions were laced with a good deal of humor. I hope that these tapes are archived somewhere.

Katmai: Al Chidester helping Bob Smith lecture

Katmai: Walt Cunningham and Alan Bean

Katmai: Al Chidester (w pipe), Bill Anders (scratching), Buzz
Aldrin, Don Wilhelms (USGS) and Bob Smith (standing with
hammer)

Katmai: Roger Chafee, Bill Anders, Dave Scott, Gene Cernan, Al Chidester

Life at the lodge was enjoyable as it was comfortable and food was plentiful. One evening I walked with Bob Smith, who it turned out was also an expert on birds, and it was fascinating to observe great horned owls and eagles and learn from him. We also were able to see someone fishing in a stream and not far upstream a bear fishing, with his paw, in the same stream.

One day we were weathered in and at some point a group of us were in one cabin. Several of the astronauts were grousing a little about geologic field work. As I had mentioned they had no problem with the geophysical instruments but they were not so comfortable with geologic mapping. At one point I said "You are all great pilots. Don't you think that the ones chosen to go to the moon will also have to be competent geologic observers?" There was instant silence. After that day it seemed there was increased interest in geology with frequent requests from many of them for additional reference material.

The astronauts were always flying NASA T-33 or T-38 jets around the country to various meetings and training sessions. I don't recall who it was that came up with the idea of providing general small scale geologic maps of the various routes their flights took but this helped their geological observations immensely.

At some point several Anchorage news people showed up at the lodge and interviewed a few astronauts. I witnessed Charlie Bassett being interviewed. During the interview he was asked if there would ever be female astronauts. He replied that undoubtedly there would be at some point but due to the physiological differences between men and women new spacesuits would have to be developed. I thought it was a great answer and later told him so. The next day the lead of the story about the astronauts was "Bassett worried about Space Brassiere". I thought we would all have to restrain Charlie as he was so livid. It was a good thing that the reporter was back in Anchorage.

I do recall the last day when we were loading all the equipment for transport having Buzz Aldrin generously pitching in and helping us load the helicopter. I was among the last to leave and the helicopter deposited me at the King Salmon airport. I can readily recommend this type of airport transport.

I flew to Anchorage in a fairly small prop plane. It had about ten seats in the back part of the cabin which was divided in half by a curtain. It turned out that my seat mate was Fritz Wein. He and his brother had started Wein Consolidated Airlines, which we were on, the forerunner to Alaska Airlines. At one point I asked Fritz what was in front of the curtain. As soon as I finished asking I heard a classic MOO! from the front. Truly, this was the frontier.

Having spent more than a week in Alaska I was scheduled to return to return to Flagstaff for several days and then travel to Iceland. However, when I returned to Flagstaff I was told by everyone that I was not going to Iceland. I had left for Alaska shortly after the birth of my first daughter and all thought that another absence was out of the question.

Medicine Lake, California

There were two three day trips a week apart, in September of that year to Medicine Lake, California. The astronauts were: William Anders, Allan Bean and Rusty Schweikart and on the second trip; Charles Bassett, Walt Cunningham, and C.C. Williams. Aaron Waters of the University of California was the leader of the first trip and Charles Anderson, USGS, led the second.

The Medicine Lake volcano is a shield volcano with extensive lava flows that range from basalt through rhyolite. The Lake is in a caldera of the volcano formed by large eruptions of andesite from vents along the caldera's rim. Just outside the eastern rim of the caldera are rhyolite and obsidian flows. The variety of volcanic features and compositional range of volcanic rocks made this an ideal training site.

The two field sessions in the Medicine Lake area were similar to the training in Alaska. Several days of geologic field lectures were followed by paired astronauts doing verbal geologic mapping via transmitting microphones. The last day was devoted to a detailed tour of the area's geology.

Our lodging at this site differed markedly from that at Brooks Lake Lodge but was equally enjoyable. We camped out using tents. It was a great fully equipped camp site thanks to the efforts of Bill Rust. Bill was a carpenter at the Branch of Astrogeology but was indispensable during our camping filed trips as he was a skilled woodsman and camp cook. We had also discovered that he was a unique character and constant source of humor. He fast became a favorite of the astronauts. It is sad that such a skilled woodsman and hunter was to die several years later in a tragic hunting accident. For some reason while hunting alone it appears that Bill chose to carry a loaded shotgun while he climbed a barbed wire fence. Apparently the gun discharged striking him in the leg and, unfortunately, he bled to death.

Camp life was pleasant and I vividly recall the many late night poker games that Al Chidester arranged. Poker by Coleman lantern in the woods in September, At one point C.C. Williams, playing cards in heavy clothing, said to Al "Hey chief, how about next year we do warm weather geology?" C.C., preferred and short for Clifton Charles, was a delightful person. He was the only Marine astronaut at the time. He mentioned to me that he was basically the Marine detachment at Houston. Each morning he would get out of bed and post the uniform of the day, shorts and tee shirt, and then return to bed. He soon arose and went to see what the uniform of the day was. We lost a dear soul when he was later killed in a plane crash.

Just before the start of the second session I was tasked with driving into Klamath Falls to pick up one of the astronauts. I don't recall which one. When I arrived at the Kingsley Field Air National Guard base, I was led to the base commander's office. He greeted me and took me to the flight control room. Shortly after we arrived we heard "Kingsley this is NASA 6" and the reply that he was cleared to land.

As he was about to touch down he gave the plane full power, shot straight up, rolled, and then back down to a smooth landing. I watched the base commander and his fight to control his temper. Fortunately, he cooled down and we went out to the air strip to greet the astronaut.

When we were in the jeep, the astronaut asked me what happened. I explained the base commander's reaction. He said "Good. That bastard wouldn't let us take off in formation the last time we were here". It was a fun ride back to camp.

Zuni Salt Lake, New Mexico

Later that month another trip was to Zuni Salt Lake, New Mexico. The astronauts were: William Anders, Allan Bean, Walt Cunningham, and Rusty Schweikart. I believe that Gene Shoemaker was the leader on that trip.

Zuni Salt Lake, a circular depression about 1 mile in diameter, is a maar type volcanic crater with complex rim material and a cinder cone within the maar. The surrounding rock consists of Cretaceous limestone and shale. Without the cinder cone the crater could be mistaken for a meteorite impact. Indeed some have suggested that such an event triggered the volcanism. This suggested occurrence was due to the fact that the Odessa Craters in Texas, Zuni Salt Lake, and Meteor Crater, Arizona lie along a line which was theorized to be the flight path of the meteor.

This was another field exercise where we did some geophysical training. Prior to the field trip we had done extensive geophysical surveying of the area and the results were published in a Branch of Astrogeology Technical Letter (Astrogeology-29). During the field exercise the astronauts conducted magnetic surveys over the crater. Afterward I showed them the results of our surveys over the Salt Lake as well as the results of our surveys over Meteor Crater. I explained the results and highlighted the differences between the geophysical survey results over this volcanic crater and a meteorite crater.

Pinacate Volcanic Field, Mexico

In November we went to northern Mexico to visit the Pinacate Volcanic Field. The astronauts were; William Anders, Walt Cunningham, Rusty Schweikart, and C.C. Williams. Dick Jahns of Stanford University was the leader.

The Pinacate Volcanic Field is a preserve in northern Mexico containing 2,000 square kilometers of basaltic rocks, shield volcano, lava tubes, maars, tuff rings, cinder cones and lava flows. It contains over 450 small volcanoes in addition to maar-caldera craters resulting from powerful steam explosions. There are three giant and six smaller maar-caldera craters as well as several tuff cones. It was an excellent and diverse area to study various volcanic features. The focus of most of the training was near the largest maar-caldera crater which is fairly unique.

Over the next year there would be several trips to the Pinacate Volcanic Field each with a different group of astronauts. In each case we camped out at the site, sans tents, but still under the care of Bill Rust. The field exercises were similar to those conducted in other locales ,i.e., a day or more of detailed discussions of the geology of the area followed by paired astronauts doing verbal geology mapping using transmitting microphones.

Pinacate: Truck with receivers for microphones, Bill Rust (USGS, back to camera), Al Chidester, Ray Zedeker (NASA).

Pinacate: Jack Schmidt walking from camp site.

I recall several stories from these trips. We traveled in our government International Harvester Travelalls, truck based utility vehicles. There were about six vehicles in our group and each had dual fuel tanks. At one point we stopped at an isolated gas station in the middle of the desert to buy fuel. I don't think I've ever seen a happier man than the station owner when we finished.

Astronaut Bill Anders (Apollo 8) had been visiting his parents and we arranged to pick him up on our way to the Pinacates. His parents had dropped him off and as we drove up, there in the middle of nowhere, was Bill standing by the road. So much for celebrity.

At the end of one trip we made a liquor run to buy a lot of booze in half gallon jugs. I was driving a Travelall towing a small house trailer that contained all of our equipment. We loaded all the bottles into the trailer and put the astronauts in the vehicle in front of mine. When we reached the border the astronauts got out of the vehicle and connected with the border agents, which made their day. They waved the other vehicles through and as I drove past the border I could hear 'Clink,Clink,Clink' sounds from the trailer. I am sure that the border agents knew what was going on but all the vehicles were U.S. Government and they were happily talking with the astronauts.

Zuni Salt Lake, New Mexico

In late December, 1965 we were back to Zuni Salt Lake with astronauts Charles Bassett, Eugene Cernan and Roger Chaffee. Gene Shoemaker once again led the trip and we did geological and geophysical exercises similar to the last time.

Grand Canyon

The next trip I was involved in was to the Grand Canyon in June,1966 Astronauts participating in this trip included Vance Brand, John Bull, Jerry Carr, Charles Duke, Joe Engle, Fred Haise, James Irwin, Don Lind, Jack Lousma, Ken Mattingly, Ed Mitchell, Bill Pogue, Stewart Roosa, Jack Swigert, Paul Weitz, and Al Worden.

The Grand canyon is an open geologic text book. The canyon walls display nearly forty major sedimentary rock layers. Also displayed are structural geologic elements such as faults. It is an ideal place for an introductory geologic field trip demonstrating sedimentary and metamorphic rocks of varying composition with varying resistance to weathering and erosion.

Ed McKee of the USGS, an expert on the Canyon, was the trip leader and it was an ideal excursion to the Canyon. Ed McKee lectured at key points as we hiked down the South Kaibab Trail. We ended at Phantom Ranch where a mule train had brought all our camping supplies. We dined at the Ranch but slept outside under the stars. Laying there I realized how different this was than mountain climbing as looking up at the high points you realize that that is ground level and you have to hike up to it in order to get home.

The next morning we hiked half way out up the Bright Angel trail and along the way Ed McKee continued his keen discussions. At Indian Gardens we were met by a mule team and lunch, We traveled by mule the rest of the way out of the Canyon. Two facts about mule trips in the Canyon. First, mules have the right of way on the trail and hikers have to step to the outside. Second, as you are riding out, mules constantly degass. I was behind Ted Foss (NASA geologist) and enveloped in a green cloud of mule gas all the way out. In our two day trip to the canyon we had secured the use of all their mules.

Grand Canyon: Ed McKee.

Grand Canyon: Facing camera Jim Irwin.

Grand Canyon: Ted Foss (NASA) on mule

Bend, Oregon

In late July,1966 we traveled to Newberry Crater area near Bend, Oregon. Astronauts participating in this trip included Vance Brand, John Bull, Jerry Carr, Charles Duke, Joe Engle, Ron Evans, Owen Garriott, Ed Givens, Fred Haise, Jim Irwin, Joe Kerwin, Don Lind, Jack Lousma, Ken Mattingly, Bruce McCandless, Ed Mitchell, Bill Pogue, Stuart Roosa, Jack Swigert, Paul Weitz, and Al Worden. Aaron Waters of the University of California was the leader on this trip.

The area of the Newberry shield volcano, with a caldera at the summit, has one of the largest collections of cinder cones, volcanic domes, lava flows and fissures in the world. The cinder cones are 200 to 400 feet high and surrounded by basalt or andesite lava flows. The northern flank of the Newberry shield has three distinct lava tubes. This area also provided a training site with a variety of volcanic features and rocks.

The trip to Bend was another one that I was delayed going to. I don't remember why. However, the USGS arranged for Jack Yockey, a friend and pilot, to fly me there in a plane the agency had leased. The most memorable part of that flight was stopping in Las Vegas to refuel. It is quite an experience to arrive there in a nice private plane. We were immediately met by a custom van and attractive hostess who put down a red carpet. She then asked what we would like to drink as the van was equipped with a complete bar. This was followed by her asking where we would like to stay (comped of course). Reluctantly, we had to defer on both accounts and said that we only stopped for fuel.

The 'camping' for this trip was in a motel and at the airport I rented a car to travel to the motel and use on the trip. Aaron did another good job leading the astronauts around the geologic wonders of the Newberry volcano. This, as had become the custom, was followed by paired astronauts doing verbal geologic mapping.

One afternoon I was driving the rental car back to the motel with three or four astronauts as passengers. On the way we suffered a flat tire. Throughout the field trips I had observed how competitive the astronauts were, For example, if one threw a rock then it soon developed into a contest as to which one could throw a rock furthest or with more accuracy. This was always done in good humor and comradeship. Thus, I was wondering how the changing of the tire would transpire. Fortunately, it developed into a coordinated, not competitive, pit stop as the group worked to see how fast they could change the tires. I stood in amazement watching several astronauts change my flat tire.

Several weeks later, back in Flagstaff, I received a letter from the rental car agency's national office stating that I had been involved in an unreported accident with the rental car. I wrote back stating that the so-called accident was only a flat tire. I also provided the names complete with titles, major, Colonel, etc., of my passengers who had changed the tire. I further noted that they could be reached at NASA's Astronaut Office in Houston, Texas. I never heard anything further from the company.

Pinacate Volcanic Field

Late November of that year found us back in the Pinacates. The astronauts were: Vance Brand, John Bull, Jerry Carr, Charles Duke, Ron Evans, Ed Gibson, Fred Haise, Jim Irwin, Don Lind, Jack Lousma, Ken Mattingly, Curtis Michel, Ed Mitchell, Bill Pogue, Harrison Schmitt, Jack Sweigert, Paul Weitz and Al Worden. Dick Jahns of Stanford University led this trip once again and as usual we camped out midst the volcanoes.

Pinacate: Bill Anders talking with Mexican observers

Apollo 1, January 27, 1967

The Apollo 1 fire was truly tragic. It impacted all of us as we had trained and known the astronauts. There also was and would be other losses in airplane accidents as the astronauts frequently traveled in their NASA jets. I recall the loss of Charlie Bassett and Eliot See in a St. Louis crash and C.C. Williams in an accident (loss of oxygen) over Florida.

The Apollo 1 tragedy had a severe impact on our geologic training program. For some time we had been working on the arrangements for a field trip to Australia. NASA canceled that trip and put travel restrictions on astronauts traveling together. However, things in the geologic training program got back to a semblance of normal within a few months.

Pinacate Volcanic Field

Trips to the Pinacates, with Dick Jahns leading the field program were now well established as was our routine of camping and running the geologic exercises. In March, 1967 we once again visited the Pinacates. Astronauts were: Joe Engle, Owen Garriott, Joe Kerwin, and Bruce McCandless.

Pinacate: USGS personnel filming and recording activities

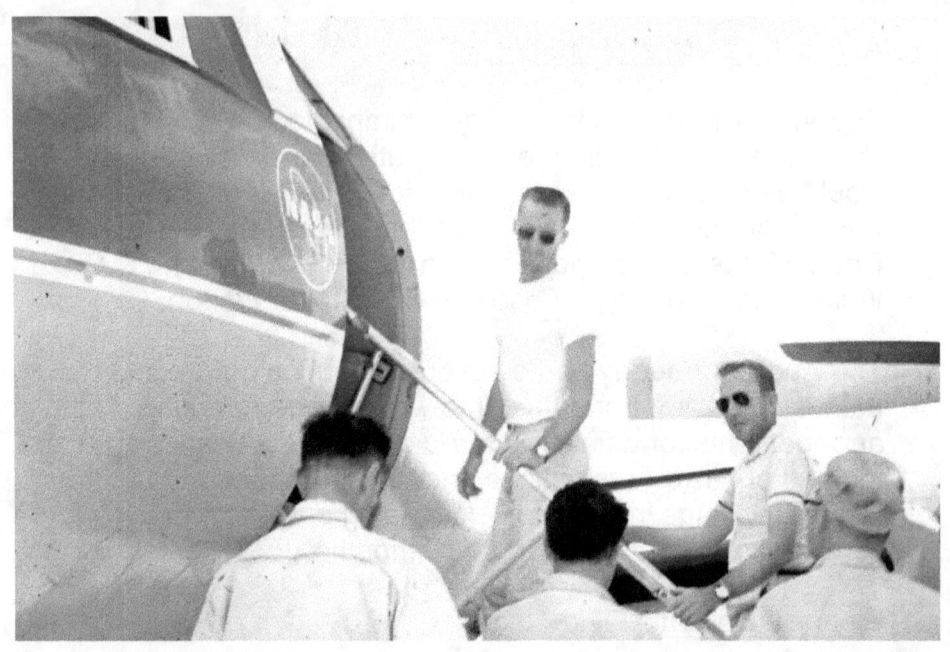

Pinacate: Astronaut Walt Cunningham and Jack Riley (NASA) departing on NASA Gulfstream.

Zuni Salt Lake, New Mexico; Hopi Buttes, Meteor Crater, Arizona

This trifecta in May, 1967 was my last trip. Astronauts were: Vance Brand, Jerry Carr, Charles Duke, Ron Evans, Owen Garriott, Fred Haise, Jim Irwin, Joe Kerwin, Don Lind, Jack Lousma, Ken Mattingly, Bruce McCandless, Curtis Michel, Ed Mitchell, Stewart Roosa, Jack Swigert, Paul Weitz, and Al Worden.

Hopi Buttes area is in a Tertiary basin of sedimentation and volcanism. Bedrock, consisting of Mesozoic sandstones and shales, is exposed in the faces of many of the buttes. Tertiary volcanics cap many of the buttes, form widespread pyroclastic deposits, protrude as dikes, and form prominent outcrops.

Gene Shoemaker was the leader on this trip that visited three diverse geologic sites. We had previously done an extensive geophysical survey of the Hopi Buttes area so during the trip I was able to give lectures on the geophysics of each site.

Of later significance to me was that Hal James, then Chief Geologist of the USGS, was on this trip. I say this because by this time I realized that I had to return to school for my Ph.D. to be able to more effectively work in the field of geophysics. On the recommendation of a colleague I had applied to Michigan State University and been accepted and awarded an assistantship.

Not long after this trip I submitted my letter of resignation. Several days later I received a call from Hal James. He said that he had my letter but asked if I had not heard of the 'In Training' program. I told him that I had but I have only worked for the USGS for several years. He replied that doesn't matter as I have been approved for the 'In Training' program and that the sound I now heard was my letter of resignation being torn up. So I went back to graduate school at full salary with tuition paid and a book allowance.

Cimarron,New Mexico: Marty Kane (hat) showing Gordon Cooper how to use a gravimeter.

EPILOGUE

I returned to Flagstaff after finishing my Ph.D. studies in June, 1969. The following month was the lunar landing of Apollo 11 and it was an especially exciting time in Flagstaff. CBS had accidentally discovered the Branch of Astrogeology and the nearby cinder field where the Branch had created, via multi-scaled explosions, a lunar landing training field complete with a LEM mock up. The Branch was essentially invaded by CBS and we all participated in interviews with George Hermann and John Hart of CBS. George Ulrich and I were interviewed by John Hart in the cinder field. After the interview one of the assistants told me that I had a great voice and asked if I had ever considered working in radio or television. I mentioned that after high school I had but was turned off by all the nepotism in the Boston area stations. Later I learned that his name was Chris Wallace, son of Mike Wallace. So, once again, I had irretrievably put my foot firmly in my mouth. At the end of the Apollo 11 coverage CBS hosted an elegant party for all of us and our families.

During the Apollo 11 landing we were all working feverishly to try to determine exactly where they had landed on the moon. I don't think that the precise landing coordinates were determined until after the mission.

Apollo 11: The CBS Trucks outside Branch of Astrogeology

Apollo 11: Jack McCauley, Bob Regan (with pipe), Thor
Karlstrom (standing), CBS assistant, Al Chidester

Apollo 11: Thor Karlstrom, Bob Regan, Jim Lovelace and George Ulrich

The success of all the Apollo missions demonstrated the effectiveness of all the geologic training, both early and later mission specific, the astronauts received. Indeed after Apollo 11, NASA awarded the USGS a Certificate of Appreciation. I found on the web a short article Ted Foss, head geologist with NASA at the time of training, wrote about the very first training series. I don't have the reference but it's title is "7. Astronaut Training in the Geosciences". I think one paragraph was quite prescient.

" During Training Series I, it was discovered that the astronauts have exceptional aptitudes and academic backgrounds to become students in the geosciences. Their strong previous training plus their great motivation have resulted in a very rapid rate of learning to a point where the students are actively and ably arguing questions of geologic fact and philosophy with their instructors. It is evident that with further training, the astronauts will be able to function as highly competent observers on early lunar landings".

I would also add that while it is well known that the astronauts underwent extensive physical testing I suspect that there were also considerable psychological evaluations. During the course of the field trips we camped and lived together for several days at a time. I have never encountered such an amiable group of individuals. There was never a problem, harsh word or discomfort in dealing with any of them.

I was only peripherally involved in the Apollo Program but I was always struck by the dedication and commitment of everyone that was involved. Neil Armstrong's foot on the Moon was the culmination of year of work by thousands of skilled people. It is a shame that at the end of the Apollo Program that this pool of talent was essentially dissolved. I always thought that NASA should have become a national problem solving agency and use the people and approach of the Apollo Program to address such issues as mass transportation, renewable energy, etc.

I would like to close by discussing Gene Shoemaker without whom the Branch of Astrogeology and the Astronaut Training Project would not have existed. I had the privilege of working with Gene on several projects. In early 1970 he was greatly concerned about meteor and asteroid impact on the Earth. At the time he was a lone voice crying in the wilderness.

After some research we set up an infrasonic (below sound) detector on the mesa behind the Branch offices. We kept hitting dead ends in our efforts to obtain more information on using infrasonics in this way. This, of course, was long before the existence of the internet. I was on vacation away from Flagstaff when I received a call from Gene saying that I had to meet him in Washington, D.C. as soon as possible. I explained that I was not near a major city and he said "Charter a plane if you have to but get there as soon as you can". So, I did, and met Gene the next day in Washington at the Cosmos Club, where he was a member.

I don't remember exactly where we went but wherever it was we were met by armed military personnel who escorted us to a meeting room. At the meeting we were told that we were intruding into National Security issues as infrasonics was being used in missile detection. It all made sense as the signature of a missile re-entering the atmosphere would be the same as a meteor. That, effectively ended our little project but not Gene's lifelong efforts to raise awareness of potential meteor or asteroid impact. He certainly was successful as we now routinely hear about such near misses on national news programs.

Gene gave a great amount to the lunar program and planetary geology. It was fitting that he was afforded his lifelong dream of going to the Moon when the Lunar Prospector carried some of his cremains to the Moon after his 1997 accidental death.

NATIONAL AERONAUTICS AND SPACE ADMINISTRATION
MANNED SPACECRAFT CENTER
Presents this

CERTIFICATE OF APPRECIATION

to

U.S. Geological Survey

For outstanding contributions to the Apollo Program. The geological training and lunar mapping services provided to the Apollo astronauts were in large measure responsible for the successful accomplishment of the scientific objectives of the world's first manned lunar landing mission in July 1969.

Signed and sealed at Houston, Texas
this month of October
Nineteen hundred and sixty-nine

DIRECTOR, MSC

Certificate of Appreciation awarded to the USGS after the successful Apollo 11 lunar landing

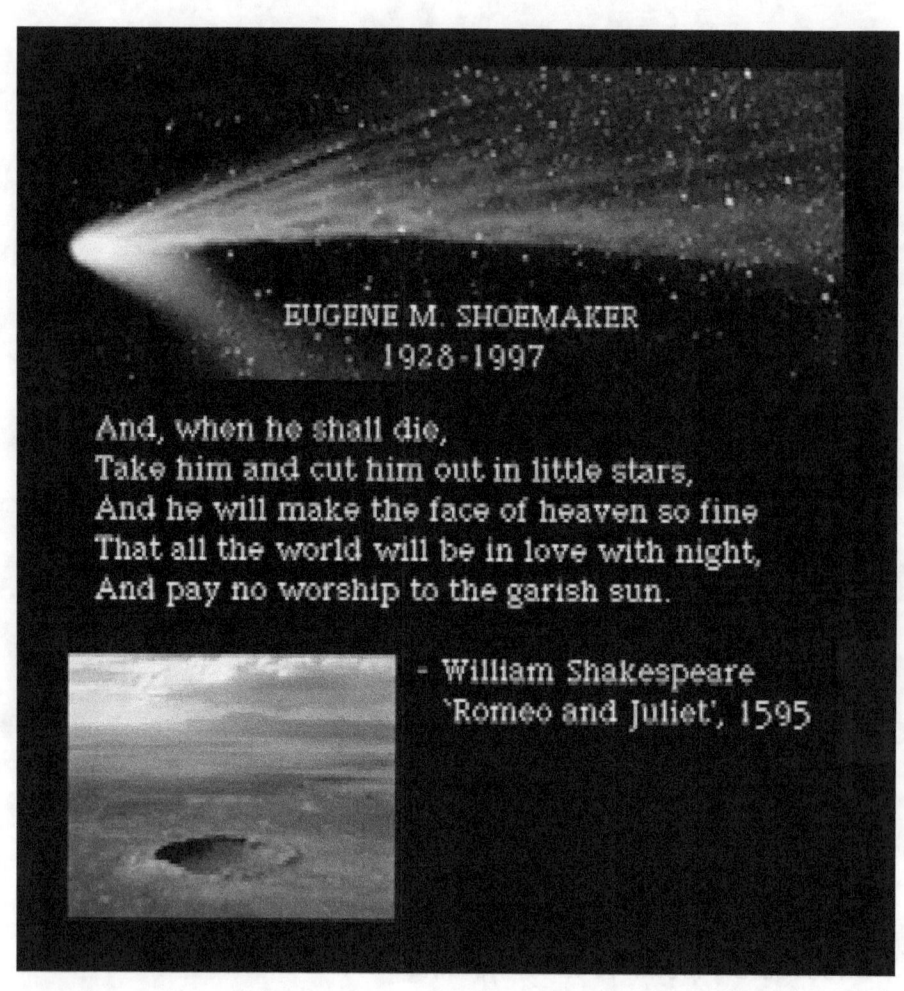

Brass foil on capsule carrying Shoemaker's cremains on
Lunar Prospector spacecraft

BIBLIOGRAPHY

To a Rocky Moon: A Geologist's History of Lunar Exploration, 1993, Don E. Wilhelms, University of Arizona Press

The U.S. Geological Survey, Branch of Astrogeology-A Chronology of Activities From Inception to the End of Project Apollo (1960-1973), 2005, Gerald G. Shaber, U.S. Geological Survey Open File Report 2005-1190

www.ingramcontent.com/pod-product-compliance
Lightning Source LLC
Chambersburg PA
CBHW070922180526
45168CB00005B/2113